YOUR KNOWLEDGE HAS VALUE

- We will publish your bachelor's and
  master's thesis, essays and papers

- Your own eBook and book -
  sold worldwide in all relevant shops

- Earn money with each sale

Upload your text at www.GRIN.com
and publish for free

Bibliographic information published by the German National Library:

The German National Library lists this publication in the National Bibliography;
detailed bibliographic data are available on the Internet at http://dnb.dnb.de .

Imprint:

Copyright © 2018 GRIN Verlag
Print and binding: Books on Demand GmbH, Norderstedt Germany
ISBN: 9783668864399

This book at GRIN:

https://www.grin.com/document/453091

Danian Singh, Kelera Railoa, Lionel Joseph

Golden Rice. An Alternative Intervention to Combat Vitamin A Deficiency

GRIN Verlag

A Review Article

Golden Rice – An Alternative Intervention to Combat Vitamin A Deficiency

Danian Singh, Kelera Railoa, Lionel Joseph.

Abstract

Vitamin A Deficiency (VAD) is a widely recognized nutritional disease affecting the world's poorest developing nations. VAD is a common affliction in Asian countries where rice is a staple food. Polished or refined rice lack essential nutrients such as carotenoids exhibiting pro-vitamin A-activity. Thus, reliance on rice as a primary food staple in low income populations contributes to vitamin A deficiency. To combat this health issue, scientists have developed Golden Rice which has been genetically engineered to contain beta-carotene in the endosperm of the grain. While this alternative solution is projected to solve the related health costs of VAD significantly there are conflicting opinions surrounding the use and effectiveness of genetically engineered crops. This review attempts to examine both ends of the spectrum in order to gauge whether or not Golden Rice could be perceived as a sustainable solution to Vitamin A deficiency.

Keywords: vitamin A deficiency, carotenoids, pro-vitamin A, Golden Rice, beta-carotene, endosperm.

Introduction

As the world progresses through this modern age, new technology and innovations are made with each passing day. The advent of such technologies and innovation has cultivated and paved the way for advances into the science of *genetic engineering*. Genetic engineering techniques involve inserting genetic material from one organism into the genome of another organism and through this leading-edge process scientists are able to manipulate and modify the genes of animals, bacteria and plants to create organisms possessing much more beneficial or desirable traits resulting in what we now know as *Genetically Modified Organisms*(GMOs)(Albaugh, 2016). These types of organisms obviously do not exist naturally in the environment and are created in developed nations which have high end equipment or simply have the financial backing to do so.

New varieties of food crops such as tomatoes, corn and wheat can be produced through genetic engineering and they are probably the most common examples of GMOs. These crops have been engineered to possess valuable characteristics such as increased vitamin content, nutritive value, better taste, overall increased yield production at low cost and resistance to various pests and diseases (Mymy, 2003).

Additionally, these crops are engineered to combat a bigger threat - food security or its lack thereof which is the leading cause of malnutrition in poor developing countries. As the human population continues to grow, so does the demand for necessities such as food. This direct correlation between human population growth and demand for food has become a persistent issue across the world's developing and poverty-ridden countries such as parts of Africa, Asia and India (Mymy,2003. Stein et. al, 2007). In these countries it is commonplace for majority of its people to lack access to decent quality food containing most if not all the essential components (vitamins and nutrients) that consist a healthy diet(Albaugh, 2016). Micronutrients are often neglected, as these are required in just minute quantities. These micronutrients include iron, zinc, iodine, cobalt, molybdenum, etcetera and not forgetting the crucial vitamins (Bollineni et al., 2014; UNICEF, 2002). The lack of a particular vitamin or mineral can lead to health conditions or nutritional deficiencies. A prevalent example of this in poor societies is Vitamin A Deficiency (VAD)which has afflicted about 19 million pregnant women and 200 million children under the age of 10 years (Dubock, 2014). Prolonged lack of vitamin A in the diet can lead to serious health issues such as blindness, a weakened immune system and cause keratinization of the urinary tracts (Nestle, 2001).

For the past two decades or so there has been a concerted effort to reduce VAD in developing countries. Notable strategies include food fortification, supplementation and dietary education programmes. Enriching major staple foods with pro-vitamin A through plant breeding has also been adopted as a complementary approach (Zimmerman et. al, 2002).Traditional breeding programmes have also identified several crop species, such as maize and sweet potato with high beta-carotene contents but because beta-carotene does not occur naturally in the endosperm of rice, the use of biotechnology is required (Bouis, 2000 in Zimmerman et. al, 2002).

Therefore, in response to reducing this problem of Vitamin A deficiency, scientists developed a genetically engineered rice variety branded as *Golden Rice.* This alternative solution to a problem that has so far claimed the lives of millions of women and children has not however been welcomed with open arms. Since knowledge regarding the long term effects of these GMOs on other organisms as well as humans is quite unknown, there still exists conflicting opinions on the consumption and benefits of genetically engineered foods such as Golden Rice (Qaim and Kouser, 2013).

This literature review therefore focuses and explores the potential impact of Golden Rice in combating Vitamin A Deficiency. It will also examine the various concerns, hazards and risks associated with the consumption of GMOs. All literature intended for this review have been extracted from Google Scholar following careful selection based on each paper's relevance to the topic and significance of information it contained. A thorough analysis and adequate understanding of key points and findings was an essential component to assimilating complex ideas expressed in the literature in order to provide a logical review.

The Development of Golden Rice

Rice is one of the most widely consumed foods particularly so in Asian countries. The part of the rice, which is actually edible, is the endosperm. While rice may be a staple food for many populations the overall essential nutrient content is very low (Dubock, 2014). Dawe et.al (2002) reports that in many parts of Asia including India, Bangladesh, Nepal, Indonesia, the Philippines, Vietnam, Cambodia, and Laos with low income and rice based diets, instances of VAD have been reported to be a significant health problem. Rising out of this health predicament, the idea of developing a rice variety genetically engineered to produce β- carotene as a solution to combat vitamin A deficiency was borne.

The endosperm of rice, as previously stated, does not naturally synthesize β- carotene however it produces geranylgeranyldiphosphate (GGPP) which is a predecessor of β- carotene (Beyer *et al*, 2002). Therefore, due to this property the use genetic engineering was required to create a special breed of rice whose endosperm would produce β- carotene. For this process,4 enzymes were needed: phytoene synthase, lycopene β-cyclase, phytoenedesaturase, and β- carotene desaturase (Mymy, 2003). These enzymes were first identified and then extracted from various bacteria and plants. It was in the early 1990s when Ingo Potrykus of Switzerland and Peter Beyer of the University of Freiburg in Germany, proposed the idea of a project to genetically engineer the pro-vitamin-A pathway into rice endosperm to the Rockefeller Biotechnology Program in New York. However from the moment this idea was first conceived, it took about 10 years following a barrage of hurdles preventing the use of genetically engineered products for public use and consumption that the actual experimentation to produce golden rice began. In the year 2000 a group of scientists were able to use agrobacterium to alter the entire β- carotene biosynthesis pathway consisting of the three genes on three vectors into rice endosperm (Beyer *et al*, 2002). This resulted in the endosperm being yellow in color due to β- carotene hence the name 'Golden Rice' was aptly coined (Mymy, 2003).

The development and success of this process showed that it was possible to influence a biochemical pathway and that too completely. This process was further developed by (Beyer *et al*, 2002) whereby the β- carotene biosynthesis pathway was transformed by either single or co-transformation of cDNA. This was carried out by using 1 gene*Carotene desaturase* from bacterium *Erwiniauredovora.* and 2 genes - *phytoene synthase*and*lycopene β-cyclase*from a daffodil *Narcissus psuedonarcissus*(Beyer *et al*, 2002). The process allowed the actual creation of golden rice and the development of this particular strain containing the highest amount of β- carotene which consequently has a direct effect on and determines the amount of vitamin A present. This final product contained about 35µg β- carotene/g of dry rice – a considerably sufficient quantity to counter vitamin A deficiency (Albaugh, 2016).

Vitamin A Deficiency

Vitamin A, also referred to as retinol, is a lipid soluble organic molecule manufactured in the body from pro-vitamin A carotenoids (Bollineni et al., 2014). Prior to its massive importance Vitamin A deficiency (VAD) is a major problem in large parts of the developing world and approximately 26 heavily populated regions such as Africa, Asia and Lain America (Bollineni et al., 2014)and is caused by insufficient consumption of foods containing vitamin A and beta-carotene (which is metabolized into vitamin A). As claimed by the World Health Organization (WHO) (WHO, 1997), close to 250 million preschoolers are Vitamin A deficient and an estimated 250,000 to 500,000 victims are becoming blind each year and 50% of the population dies under an year of losing sight (Chouhan, 1977). Preformed vitamin A can be naturally sourced from animal products while beta-carotene can be found in dark green, yellow, and orange plants. Precursor forms, in turn, are again found in many fruits and vegetables, as well as in yeast and some species of bacteria and fungi. The precursors of VA are known as carotenoids with the most well-known carotenoid being beta-carotene, which the human body can convert to the active form of VA (Zimmerman et. al, 2002). While VAD is commonly, a result of malnutrition it can also be due to abnormalities in the intestinal absorption of retinol or carotenoid. Vitamin A is a fat-soluble vitamin, which is stored primarily in the liver in the form of retinol derivatives. The vitamin carries out several important bodily functions the principal being sight and general eye health promoting good vision.

In addition to its role in reproduction and lactation it also helps in the development and maintenance of healthy teeth, skeletal and soft tissue, mucous membranes, and skin (Sommer and West, 1996). Vitamin A occurs in two forms: the active (preformed) and the precursor form. Preformed VA is found only in animal products; the most frequent active VA is retinol.

VAD is unique in that of all the numerous nutrient deficicney afflicting poor populations, it has well defined and distinctive effects. When a person does not consume enough vitamin A and beta-carotene for extended periods of time, the most recognizable symptoms include night blindness or nictalopia where a person is not able to see clearly in dim light or even upon exposure to bright light, Bitot's spot (foamy patches on the surface of the eye), xeropthalmia, and in extreme cases, blindness (National Academy of Sciences, 2001).Among young children VAD has also been associated with stunted growth, increased mortality, and vulnerability to infection, particularly measles and diarrhea (Sommer, 1997 in Dawe et.al. 2002).

An estimated 250,000 to 500,000 VA-deficient children go blind every year(Zimmerman et. al, 2002).VAD symptoms are not limited to acute eye symptoms but also weaken the immune system, increasing susceptibility to infectious diseases and infant mortality rates. In affected adults, the implications are often also serious especially for pregnant and lactating women. An estimated 600,000 women die from childbirth-related causes each year, many of them from complications, which could have been reduced or avoided through better provision of vitamin A (Sommer and West, 1996). The poor are the most affected populations mainly due to their lack of purchasing power and public health awareness, which restricts them to diets that are predominated by low-cost staple foods such as rice which is lacking in nutritive value.

Interventions to Combat Vitamin A Deficiency

Overall, five different interventions to combat micronutrient malnutrition such as VAD can be differentiated. These include: pharmaceutical supplementation, industrial fortification, biofortification, dietary diversification and supporting public health measures. Most micronutrient deficiencies can be effectively addressed through dietary diversification whereas in areas where the traditional diet lacks a specific nutrient, such as Vitamin A for instance - fortification strategies are wanted. Countries that lean more towards "greener" methods preferably opt for food-based approaches to fulfilling dietary requirements and continue to receive strong support (WHO/FAO, 1996; FAO/WHO, 2002). These strategies are shortly discussed below.

Pharmaceutical supplementation

The use of supplements is considered an effective, short term, technical approach to deliver micronutrients directly via syrups or pills. While supplements are able to deliver an adequate level of protection against micronutrient deficiencies it does not tackle the root of dietary deficiencies, which is insufficient and poor quality food. The major drawback of this approach also arises in the cases whereby broader groups of society are affected, or say for instance if the supplementation programme is to reach target groups in remote or isolated communities, the resources needed(financial, administrative and human) will become substantial. So much so that it will surpass the costs of the capsules therefore making this approach increasingly ineffective, and unsustainable over the long-run. Additionally some target groups may not accept supplements and, in the case of supplements that need to be taken regularly, reaching an agreement or at best, a compromise will become an issue. A limitation is that supplementation needs to be delivered to the public via public health services, but the needy people in remote rural regions are rarely having public health centers to make this possible (De Steur et al., 2017).

Industrial fortification

Fortification works on the basis of utilizing widely available and commonly consumed foods to deliver micronutrients to populations at risk. The most commonly recognized example of this strategy is the fortification of salt with iodine yielding what we know as iodized salt. Ensuring that fortification strategies are successful requires integrated processing facilities, strong mechanisms for quality control, effective social marketing and public education strategies (Nantel and Tontisirin, 2002; Uauy et al., 2002 in Kennedy et. al. 2003).

Biofortification

Similar to industrial fortification, biofortification efforts have the objective to also increase the micronutrient content in foods consumed by deficient populations except in this case fortification is not carried out during food processing but through plant breeding. Hence, unlike with industrial fortification, biofortification is cost-effective. This is mainly due to low monitoring costs and the lack recurrent costs for buying the fortificant as it involves just one initial investment, which is the development of a biofortified staple crop that is adaptable and can be utilized globally. Likewise, once farmers grow these biofortified crops, the produce will continue to reach at-risk populations even if resources and public attention shift away from the problem. Another case in point with regards to beta-carotene fortification an added gain is that the risk of overdosing is removed because the human body will only convert as much beta-carotene into vitamin A as it needs. The only slight drawback would be the increased instability of beta-carotene when stored or exposed to light and heat.Moreover, Helen Keller International (HKI), a leading global health organisation in reducing VAD and resulting child mortality is promoting the production and consumption of orange-fleshed sweet potatoes (Coghlan, 2013). These orange-fleshed sweet potatoes are bred to yield higher beta-carotene in comparison to the traditionally grown sweet potatoes with minimum beta-carotene content. However, since sweet potatoes have some content of naturally occurring beta-carotene, it becomes easy for the breeders to use traditional method to increase the content. On the other hand, rice, whether cultivated or wild does not contain any level of beta-carotene. Prior to this, it becomes rather impossible to introduce beta-carotene without genetic engineering and since sweet potato breeding does not involve any genetic modification, there are no complaints from anti-GM activists about it. On the other hand, although this strategy might have its pros, an important fall back would be its high cost and inaccessibility to people residing in the remote rural regions (Bollineni et al., 2014; Mayer et al., 2008).

Dietary diversification and Behavior change

This strategy aims at increasing the intake of foods high in micronutrients or that are generally accepted as healthy. These include food groups like green leafy vegetables, orange-fleshed fruits and tubers, citrus fruits and animal source foods (Stein et. al. 2006). Diversification can be improved through intensifying (own) production, processing, and marketing of such food or through education and awareness programmes aimed at changing cooking methods and eating behavior. Dietary diversification may be generally considered the most sustainable solution in the long run, because it improves overall dietary quality instead of addressing single micronutrient deficiencies only.

General public health measures

Public health measures complement micronutrient interventions and reduce primary causes of malnutrition while also improving the overall standard of living. These measures include but are not limited to the provision of clean drinking water, improvements of personal hygiene, de-worming and better access to health-care services (Stein et.al.2006). These interventions can be ranked according to their effectiveness and sustainability. An integrated approach, using a mixture of direct and indirect interventions and public health measures, with the inclusion of education and awareness campaigns have proven successful in reducing micronutrient malnutrition (Underwood, 1999).

Effectiveness of Golden Rice

Effectiveness or the usefulness of a particular product is a key factor in determining whether a product is worth the effort, research and resources that have been invested into its development. A product that is unable to fulfill the purpose for which it was intended is any inventor or innovator's worst nightmare. Fortunately enough for the scientists and genetic engineers behind the Golden Rice project this product has met with resounding success albeit an almost decade long delay. The central motivation behind the creation of golden rice was to deliver a solution towards the treatment and prevention of vitamin A deficiency to the most afflicted populations across poor and developing societies. There have been numerous studies conducted which show concrete evidence of the efficiency of golden rice in combating vitamin A deficiency when tested on humans. One such study was carried out in China on a group of 68 elementary school students aged between 6 to 8 (Tang *et al*, 2012). The subjects were divided into three groups and each group was given a different source of vitamin A. The first group was given spinach whereas the second group was given golden rice. The third group was given an oil capsule, which contained pure β-carotene. The children's diet was carefully monitored at home and also at school so that there was no outside source of vitamin A intake apart from the ones provided by the researchers. Blood samples were extracted every 3rd day for 21 days to examine the levels of vitamins. The results of this research indicated that golden rice, β-carotene in the oil capsule together with spinach all have the ability to provide vitamin A. Out of these 3 sources golden rice was as effective as pure β-carotene. Overall β-carotene in oil capsule and golden rice were much more effective then spinach.

A second study was conducted in Thailand to determine the value of vitamin A in humans using dietary golden rice (Tang *et al*, 2009). In this research golden rice was fed to 5 volunteers out of whom 2 were men and 3 women. The subjects were carefully examined prior to experimentation to see if they were healthy and had not taken any vitamin A supplements within the past month. Blood samples from these volunteers were collected for 36 days. The subjects were given a reference dose of retinly acetate oil. This was given one week preceding the golden rice ingestion. The final results indicated that 3.8µg of golden rice can provide about 1µg of retinal which is a type of vitamin A. Overall this research concluded that golden rice is a very efficient source of vitamin A and about only 25% is needed of the daily recommended allowance to prevent blindness and death which is associated with vitamin A deficiency.

While golden rice has the scientific backing of its ability to provide vitamin A there are various other issues that hinder its growth globally. Firstly the bioavailability of β- carotene is very low since there are other dietary requirements for its breakdown to vitamin A. These requirements are dietary fat, adequate proteins and most importantly a healthy functional digestive system (Nestle, 2001). As previously mentioned, in countries where vitamin A deficiency is a public health issue, many people have exhibited the lacking of these requirements (Kouser and Qaim, 2013). This can lead to golden rice having little to no effect on these target groups as the issue lies not in a dietary deficiency but is due to the body's inability to process vitamin A.

Hazards and Concerns on the Usage of Genetically Modified Plants

The production and usage of GMOs such as golden rice to fight against public health issues has garnered a lot of public attention and support since the inception of the idea. But despite genetically engineered foods being scientifically proven to be capable of delivering essential macro and micronutrients to benefit the human diet they continue to be a subject of controversy in the face of genetic engineering (GE) regulation. Critics of the use of genetically engineered products as a public good have highlighted several concerns and possible hazards posed by such products.

Firstly in a situation where the products of various genetically modified plants are consumed by animals and humans, there always is a fear that these plants may create various health issues. Gene transfer from plant to microbes is one of the potential risks in which researchers and the public both are concerned because it can lead to antibiotic resistance. It is thought that the consumption of various genetically modified foods which contain antibiotic resistance genes may lead to microbes becoming resistant to various types of antibiotic therapies therefore increasing the difficulty with which bacterial and viral infections can be treated. To illustrate this scenario, the Npt II gene encodes for neomycin phosphotransferase, which is resistant to kanamycin and neomycin. The gene could be transferred from genetically modified food such as tomatoes to guts of animals and humans, which contain microbes, and also to the pathogens present in the environment (Verma, 2013).

Another apparent health concern is that genetically modified plants may have the ability to generate new pathogenic organisms. It is speculated that various microbial strains, which are non-pests, may attain pathogenic qualities through gene flow from genetically modified plants (Mayer, 2005). Also the new virus resistant plants may allow infection of new hosts. Various plant pathologists have hypothesized that the growth of new viruses resistant plants can allow virus to contaminate new hosts and these plants may also allow the formation of new viruses through exchange of genetic materials (Verma, 2013).

Embedded in all the possible health issues that may arise out of the utilization of genetically modified crops is the general skepticism of people about the nutritional value of such foods. Golden rice therefore has been at the forefront of this skepticism (Dubock, 2014). Various activist groups actively claim that the degradation of golden rice is very fast and that only a small proportion of vitamin A left over after cooking (Verma, 2013). Additionally, golden rice in its ability to cure vitamin A deficiency has dragged away attention from other nutrient deficiencies, which have taken a back seat as researchers, and governments focus on vitamin A. In this instance, the development of golden rice draws nearly all research and financial resource that could also be effectively utilized to treat other equally troubling deficiencies (Mayer, 2005). The usage of alternative food products other than golden rice is another means to combat vitamin A deficiency and this is an argument often pointed out by activists from organizations such as Greenpeace and Friends of the Earth who have emphasized the use of greener and more environmentally friendly alternatives such carrot, green leafy vegetables, mangoes and jackfruit against vitamin A deficiency rather than consuming golden rice (Potrykus, 2001). However scientists and experts have mentioned that the availability of these alternatives are a major issue in countries which are poverty ridden and have high amounts of a Vitamin A deficiency. Also worth noting is that a varied diet is far beyond the reach of many poor people so they mainly rely on a few foods for sustenance even if these staple foods may not in fact provide them with all the nutrients and vitamins required for a healthy diet(Albaugh, 2016).

Moreover, apart from health concerns there are various environmental hazards and concerns regarding the usage of genetically modified foods. There have been several debates on the introduction of genetically modified plants commercially in many parts of the world and the impacts it will have on the environment. Firstly breeding of genetically modified plants with their wild relatives is a potential risk. This risk stems from the possibility that modified plants may cross breed with their compatible wild relatives, made possible through wind pollination and seed dispersal (Verma, 2013). Interbreeding will have adverse effects on environment balance due to the creation of hybrids that are stronger, better yielding and highly resistant to pest and disease. For example genetically modified virus resistant squash became commercial in1994 and exhibited the transfer of its virus resistance trait gene to its wild squash relatives (which are weeds) thereby decreasing its overall value to farmers (Kouser and Qaim, 2013).

Interbreeding is also an issue of worry among farmers and environmentalists as these genetically modified plants proliferatein nature. This will allow the natural environments, which do not have GMOs to be contaminated, and this will affect future generations of plants in an uncontrollable way (Qaim and Kouser, 2013). Insects that are pests can gain resistance to pesticides, which will create another environmental issue. It is probable that various disease resistant genetically modified plants can drive the evolution of insect pests letting them becoming resistant as well(Lu and Snow, 2005). For example, *Bt* crops have the ability to develop resistance to *Bt*biopesticide which is used by organic farmers to control pests, however to this date there are no reported cases where insects have become resistant to *Bt* crops under normal conditions (Verma, 2013).

However, *Bt* resistant insects such as the bollworm have been observed in places where *Bt* pesticide has been used (Verma, 2013). This poses a serious threat where farmers may resort to the usage of less environmentally friendly pesticides.

Loss of biodiversity in plants is another issue but one of great importance. Major losses in biodiversity have resulted mainly due to industrialization, urbanization and the increased use of genetically modified high yielding plants (Qaim and Kouser, 2013). It is anticipated that genetically modified plants will further harm natural biodiversity due to the encouraged use of these plants. This is mainly because modified plants may favor monocultures, where plants of a single type are grown yielding the same products and are suitable to one type of condition as well (Dubock, 2014). This will further cause the transformation of natural ecosystems into agricultural fields for the planting of these crops (Verma, 2013).

Role of ethics in decision-making

Global introduction of global rice is definitely not an easy task and a collective decision-making is required. Ethics come into play, as most people do not completely believe in GM crops as it is often thought that adopting golden rice as a solution against VAD could backfire, as it is not natural but artificially created. Genetic modification remains controversial and due to proper regulatory structure and setbacks have hindered golden rice from reaching people who need it urgently (Eisenstein, 2014). Before commercializing golden rice, certain code of ethics and standards needs to be set that the engineers must abide by when genetically modifying rice. In accordance to (Judge, 2013), it is essential to think about the consumers' health as they are the ones who will consume these foods and if a product is not safe, it should not be made available in the markets. Also, it is morally incorrect and unethical to spread false information to the public about something just for the sake of making sales and profit.

Furthermore, Golden rice has been there for more than a decade but is still not commercialized in majority parts and a major factor for this delay is people's fear of GM foods. Some argue (Judge, 2013) that it is not ethically correct to ignore consumer choice on what they want but it is also true that many people oppose GM foods without any strong evidence that golden rice is bad. There are no adverse short-term side effects observed so far but tests need to be conducted on a long-term basis as well (Judge, 2013).

In some countries, golden rice could be accepted more easily. For instance, for some Asian countries such as China, Japan and others rice is considered as a major meal and the people are relying mostly on rice. For instance, in Philippines, a meal is not considered a meal unless and until rice is included (Eisenstein, 2014). So, for those countries, it could be easier to use golden rice without much ethical conundrum. Price also plays a crucial role, since extra task is needed to formulate golden rice means its price would be slightly higher than the price of other brands of non-GM rice. In poor developing countries, people would just choose cheaper brands and hardly anyone would want to purchase golden rice due to its price and bringing in golden rice in those countries could be a failure. An interesting point made in (Judge, 2013; Eisenstein, 2014) is that the sole reason of placing golden rice on the market is not for the sake of helping people living in remote undeveloped countries but it is for making profit by agricultural companies.

A study undertaken in China (Jin et al., 2014) based on the consumer acceptance and willingness to pay for genetically modified rice in China. A total of 994 respondents were assessed and 73% of the population opted for non-GM rice in comparison to GM rice and the respondents suggested that a discount of 68% would be crucial to make GM rice more competitive in the market. However, on the contrary, a study by Zhong et al(2003) indicated that not many people in China are informed about GM crops; regardless of this, 40% are ready to invest in GM crops. Another research (Huang et al., 2008) asserts that 60% of the people are ready to purchase GM food, 20% would not buy and 20% would buy with price discount on these GM food items. Furthermore, as mentioned in (De Steur et al., 2013; Jin et al., 2014) there has been a vast decline in consumer acceptance of GM food from 80% in 2005 to 40% in 2010. Golden rice was to be introduced in India by 2002 but it was not due to ethical issues, but calculations in (Wesseler & Zilberman, 2014) shows that with consumption of golden rice 204,000 life years can be saved per year and a delay in introducing this crop might have caused losses of more than 1,424,680 life years for India. In accordance to (Eisenstein, 2014), the scientific problems existing can be solved over the years upon improvements and tests, but peoples fear about GM crops and ethical issues may be a bigger problem to overcome. For example, European and North American activists have raised concerns and doubts of environmental and health influences of artificially created GM golden rice.

Majority of the respondents in studies that show positive attitude towards GM crops are often not well informed on the subject and are mostly uneducated (De Steur et al., 2017). Respondents that are more educated always tend to bring in ethical viewpoints to debate and bring in the negative side of GM crops. To some extend it is correct, because introduction of a GM crop needs to be looked from all angles and the impacts and cost-effectiveness of GM crops needs to be known through ex post analysis as well (De Steur et al., 2017). Thus, another important thing that needs to be considered is that despite having positive effect on nutrition, social welfare and overall public health, slight spread of negative information on GM crops can cause repellence at the consumer level.

Recommendations

If seen and observed carefully different studies are portraying different ideas and believe. Some are supporting the idea that golden rice needs to be made widely available in the market to combat VAD while some are making assumptions that golden rice could have negative impact on the human health and environment. There has been no short-term side effects observed so far and if tests show that there are no such side effects for long-term studies as well then everyone needs to agree that consumption of golden rice instead of the normal rice would play an important role in fighting against VAD(Judge, 2013).

Global standards and code of ethics need to be formulated so that proper set of measures are taken by engineers when genetically modifying rice to make golden rice. The standards and code of ethics need to be followed by all the nations planning to develop GM crops to ensure proper and efficient development. This will also ensure honest practice by producers and build confidence in consumers to invest in golden rice.

The price of golden rice will certainly be higher compared to ordinary rice because extra energy and effort are required to produce it. This higher cost might discourage consumers to choose this brand and it will hardly reach people living in remote rural regions. Per say introduction of this crop was to reduce VAD and many people suffering from this reside in rural areas of undeveloped countries. These people are poor and will not be able to afford this GM rice; instead, they might opt for cheaper product. However, if government intervenes and provides incentives to make its price lower through discounting, then many people would afford and prefer it.

Furthermore, most people are in fear of GM food crops although they might not really know about it. Slight incorrect information about a product can spread rapidly and make consumers lose complete faith. Therefore, to avoid this confusion and mislead information proper education needs to be provided to the consumers. Advertisements in television, radio and newspapers will be useful and also organizing awareness campaigns to inform the general public about VAD and how golden rice can combat it.

VAD is a serious problem and fighting it is a global aim. If proper strategies and frameworks were developed then introducing golden rice in markets would no longer be a delayed action. In addition, government action will be required to provide incentives to lower the price of GM rice so that consumers show interest in purchasing those brands. Price is a very crucial factor as it can play with the emotions of the consumers. For instance, if the product is far too expensive then consumption rate would decline and markets would stop the production if no profit were made. This product also needs to focus on solving VAD crisis in remote rural regions, as people living in urban centers would easily purchase it but it might be difficult to reach in interior areas. Therefore, for this, health centers could be set up in rural areas to provide golden rice even in the interior localities.

Conclusion

This paper has reviewed the utilization of Golden Rice as an alternative to combatting Vitamin A deficiency and has found this product to be an effective solution. Both major studies carried through human trials have evaluated the efficacy of golden rice and provide scientific evidence that only about 25% is necessary of the recommended intake to meet dietary vitamin A requirements. In both studies there were no negative effects on the test subjects. Additionally, there exist no current studies mentioning whether texture of golden rice is an important factor determining consumer consumption.

Various health and environmental concerns have been highlighted and while the use of genetically modified crops has shown signs of negative impacts in these sectors the magnitude of the hazards posed by these crops depends on how extensively they are used and the eventual production yielded. The short-term effects of GM crops are nearly negligible but the real extent of long-term effects are not fully understood and more studies needs to be undertaken to learn about these.

Genetic engineering of plants is a very modern technique requiring a much greater deal of human and financial resource when compared to traditional cross breeding methods. Despite the differences in the techniques employed, both methods share the same goal which is to produce larger, tastier and higher yielding plants with higher nutrient content. As the science behind genetic engineering continues to grow there is a need to have various safety mechanisms in place so that it does not pose a long-term threat to human health, the environment and ecological biodiversity. Extensive studies need to be conducted to measure the effects that these GMOs will have on human health and the environment. The concerns regarding the use of GMOs in combating public health issues and improving human health is natural as these products are relatively new and it will in fact take time for people to accept new products.

Golden rice is only an additional alternative or solution to the other existing strategies such as fortification, supplements and dietary diversification. However, given that Golden Rice is cost effective and much more sustainable compared to other interventions, it is probably worthwhile to consider it as a permanent solution to micronutrient malnutrition. On a last note, the setbacks that have hindered the introduction of golden rice are rather unfortunate, but in years to come experts and researchers would come up with a more convincing solution to make golden rice commercially available globally.

References

Albaugh, M. (2016). Golden Rice: Effectiveness and Safety, A Literature Review. Honors Research Projects. Paper 382.

Beyer, P., Babili, S., Ye, X., Lucca, P., Schaub, P., Schaub, R., &Potrykus, I. (2002).Golden Rice: Introducing the β-carotene Biosynthesis Pathway into Rice Endosperm by Genetic Engineering to Defeat Vitamin A Deficiency.*The Journal of Nutrition, 20*(5), 506-510.

Bollineni, H., Krishnan, S., Sundaram, R., Sudhakar, D., Prabhu, K., & Singh, N. et al. (2014). Marker assisted biofortification of rice with pro-vitamin A using transgenic Golden Rice®lines: progress and prospects. *Indian Journal Of Genetics And Plant Breeding (The), 74*(4s), 624. http://dx.doi.org/10.5958/0975-6906.2014.00901.8

Chouhan, B. (1977). Kajal eye staining for early detection of vitamin A deficiency in the field.*Tropical And Geographical Medicine, 29*, 278-282.

Coghlan, A. (2013). Golden Rice creator wants to live to see it save lives. *New Scientist,220*(2941), 30-31. http://dx.doi.org/10.1016/s0262-4079(13)62588-9

De Steur, H., Buysse, J., Feng, S., &Gellynck, X. (2013). Role of Information on Consumers' Willingness-to-pay for Genetically-modified Rice with Health Benefits: An Application to China. *Asian Economic Journal, 27*(4), 391-408. http://dx.doi.org/10.1111/asej.12020

De Steur, H., Demont, M., Gellynck, X., & Stein, A. (2017).The social and economic impact of biofortification through genetic modification. *Current Opinion In Biotechnology, 44*, 161-168. http://dx.doi.org/10.1016/j.copbio.2017.01.012

Dubock, A. (2014). The Present Status of Golden Rice.*JournalOfHuazhong Agricultural University, 33*(6), 70-83.

Eisenstein, M. (2014).Against the grain. *Nature Biotechnology, 514*, 55-57.

Huang, J., Hu, R., Rozelle, S., & Pray, C. (2008). Genetically Modified Rice, Yields, and Pesticides: Assessing Farm-Level Productivity Effects in China. *Economic Development And Cultural Change,56*(2), 241-263. http://dx.doi.org/10.1086/522898

Jin, J., Wailes, E., Dixon, B., Nayga, R., &Zheng, Z. (2014). Consumer acceptance and willingness to pay for genetically modified rice in China. In *Agricultural & Applied Economics Association's 2014 AAEA Annual Meeting*. Minneapolis.

Judge, L. (2013). Should Golden Rice be Put on the Market?. *University Of Pittsburgh, Swanson School Of Engineering*. Retrieved from http://www.pitt.edu/~lmj36/Writing%20Assignment%203.pdf

Kennedy, G., Nantel, G., &Shetty, P. (January 01, 2003). The scourge of "hidden hunger": Global dimensions of micronutrient deficiencies. Food, Nutrition and Agriculture

Lu, B. & Snow, A. (2005).Gene Flow from Genetically Modified Rice and Its Environmental Consequences.*Bioscience*, *55*(8), 669.

Mayer, J. (2005). The Golden Rice Controversy: Useless Science or Unfounded Criticism?.*Bioscience*, *55*(9), 726.

Mayer, J., Pfeiffer, W., & Beyer, P. (2008).Biofortified crops to alleviate micronutrient malnutrition. *Current Opinion In Plant Biology*, *11*(2), 166-170. http://dx.doi.org/10.1016/j.pbi.2008.01.007

Mymy, B. (2003). Golden Rice: Genetically Modified to Reduce Vitamin A Deficiency, Benefit or Hazard?.*Nutrition Bytes*, *9*(2), 1-4.

Nantel, G. &Tontisirin, K. 2002.Policy and sustainability issues. J. Nutr., 132: 839S–844S

National Academy of Sciences, 2001.Dietary Reference Intakes for Vitamin A, Vitamin K, Arsenic, Boron, Chromium, Copper, Iodine, Iron, Manganese, Molybdenum, Nickel, Silicon, Vanadium, and Zinc.http://books.nap.edu/books/0309072794/html/86.html.

Nestle, M. (2001). Genetically Engineered "Golden" Rice Unlikely to Overcome Vitamin A Deficiency.*Journal Of The American Dietetic Association*, *101*(3), 289-290.

Potrykus, I. (2001). Golden Rice and Beyond.*PLANT PHYSIOLOGY*, *125*(3), 1157-1161.

Qaim, M. &Kouser, S. (2013). Genetically Modified Crops and Food Security.*Plos ONE*, *8*(6)

Sommer, A., 1997.Vitamin A prophylaxis. Archives of Disease in Childhood 77, 191–194.

Sommer, A., West, K.P., 1996.Vitamin A Deficiency: Health, Survival, and Vision. Oxford University Press, New York.

Stein, A. J., Sachdev, H., &Qaim, M. (2006).Potential impact and cost-effectiveness of Golden Rice. Nature Biotechnology, 24(10), 1200-1201. doi:10.1038/nbt1006-1200b

Tang, G., Hu, Y., Yin, S., Wang, Y., Dallal, G., Grusak, M., & Russell, R. (2012). -Carotene in Golden Rice is as good as -carotene in oil at providing vitamin A to children. *American Journal of Clinical Nutrition*, *96*(3), 658-664.

Tang, G., Qin, J., Dolnikowski, G., Russell, R., &Grusak, M. (2009). Golden Rice is an effective source of vitamin A. *American Journal Of Clinical Nutrition*, *89*(6), 1776-1783.

Uauy, R., Hertrampf, E. & Reddy, M. 2002. Iron fortification of foods: Overcoming technical and practical barriers. J. Nutr., 132: 849S–852S.

Underwood, B. 1999.Perspectives from micronutrient malnutrition elimination/eradication programmes. MMWR – Morbidity and Mortality Weekly Report, 48: 37–42.

UNICEF.(2002). Vitamin A global initiative. Retrieved from http://www.unicef.org/vitamina

Verma, S. (2013).Genetically Modified Plants: Public and Scientific Perceptions.*ISRN Biotechnology*, *2013*, 1-11.

Wesseler, J., &Zilberman, D. (2014).The economic power of the Golden Rice opposition.*Environment And Development Economics*, *19*(06), 724-742. http://dx.doi.org/10.1017/s1355770x1300065x

WHO. (1997). Vitamin A supplements: a guide to their use in the treatment of vitamin A deficiency and xerophthalmia. Geneva, Switzerland.

Zhong, F., Marchant, M., Ding, Y., & Lu, K. (2003). GM foods: A Nanjing case study of Chinese consumers' awareness and potential attitudes. *Agbioforum,*, *5*(4), 136-144.

Zimmermann, Roukayatou; Qaim, Matin (2002) : Projecting the benefits of golden rice in the Philippines, ZEF Discussion Papers on Development Policy, No. 51

YOUR KNOWLEDGE HAS VALUE

- We will publish your bachelor's and
  master's thesis, essays and papers

- Your own eBook and book -
  sold worldwide in all relevant shops

- Earn money with each sale

Upload your text at www.GRIN.com
and publish for free